CONTENTS

Hunters of the Tundra

Arctic wolves live in the **tundra** of North America and Greenland. The weather there is very cold. However, these wolves are good at living in their Arctic homes. They have many **adaptations** that help them.

ANIMALS OF THE TUNDRA

WILD ARCTIC WOLVES

THERESA EMMINIZER

PowerKiDS press.

New York

Published in 2022 by The Rosen Publishing Group, Inc.
29 East 21st Street, New York, NY 10010

First Edition

Portions of this work were originally authored by Maeve T. Sisk and published as *Arctic Wolves*. All new material in this edition authored by Theresa Emminizer.

Editor: Jill Keppeler
Book Design: Michael Flynn

Photo Credits: Cover, p. 1 Byrdyak/iStock/Getty Images; (series background) Stefan Sorean/Shutterstock.com; p. 5 John Knight/Moment Open/Getty Images; p. 7 (main) Cybernesco/iStock/Getty Images; p. 7 (inset) JohnDPorter/iStock/Getty Images; p. 9 www.digitaldirectphotos.com -Portfolio www.flickr.com/michel/Moment/Getty Images; p. 11 Fuse/Corbis/Getty Images; p. 13 libre de droit/iStock/Gettty Images; p. 15 Picture by Tambako the Jaguar/Moment Open/Getty Images; p. 17 Dee Carpenter Originals/Shutterstock.com; p. 19 Raimund Linke/The Images Bank/Getty Images; p. 21 Jim Cumming/Shutterstock.com.

Library of Congress Cataloging-in-Publication Data

Names: Emminizer, Theresa, author.
Title: Wild arctic wolves / Theresa Emminizer.
Description: New York : PowerKids Press, [2022] | Series: Animals of the tundra | Includes index.
Identifiers: LCCN 2020021701 | ISBN 9781725326552 (library binding) | ISBN 9781725326538 (paperback) | ISBN 9781725326545 (6 pack)
Subjects: LCSH: Gray wolf–Juvenile literature. | Tundra animals–Juvenile literature.
Classification: LCC QL737.C22 E596 2022 | DDC 599.773–dc23
LC record available at https://lccn.loc.gov/2020021701

Manufactured in the United States of America

Some of the images in this book illustrate individuals who are models. The depictions do not imply actual situations or events.

CPSIA Compliance Information: Batch #CSPK22 For Further Information contact Rosen Publishing, New York, New York at 1-800-237-9932

Changing Coats

Arctic wolves often have a white coat of fur that helps them hide in snow and ice. They can also be gray or other colors. Their fur is very thick to keep them warm.

Staying Warm

The heavy fur on an Arctic wolf's feet lets it walk in the cold tundra. Its short **muzzle** and small ears help it hold in body heat. Other kinds of wolves have a longer muzzle and bigger ears.

Predators and Prey

Arctic wolves are **predators**. Many other animals of the tundra are their **prey**. Musk oxen, caribou, Arctic hares, and other small animals are some of their favorite meals. The wolves hunt as a pack to catch bigger prey.

Pack Hunters

A pack often has about seven or eight wolves. They live, hunt, and travel together. They may travel many miles to find food. One pack can have a territory, or area it uses, of 1,000 square miles (2,590 sq km).

Fuzzy Pups

Only the **male** and **female** leaders of a wolf pack have pups. They'll have about two to seven pups each year. Newborn Arctic wolf pups are brown. They have blue eyes. The pack keeps them safe.

Growing Up

In time, a pup's fur grows lighter. Its eyes turn gold or brown. The other wolves of the pack teach the pups to hunt. When a pup turns about six months old, it starts traveling with the pack.

I'm Here!

Arctic wolves talk to each other by howling. Sometimes wolves howl to tell each other where they are. Sometimes they howl to start a hunt. They may also warn other wolves to stay away.

Dangers

Some people hunt Arctic wolves. However, the Arctic wolf doesn't usually see people in its cold home. Still, changes in weather caused by humans affect the wolves' food supply. How can we help keep these tundra animals safe?

Fancy Features

Length

3.3 to 5.9 feet (1 to 1.8 m) long, including tail

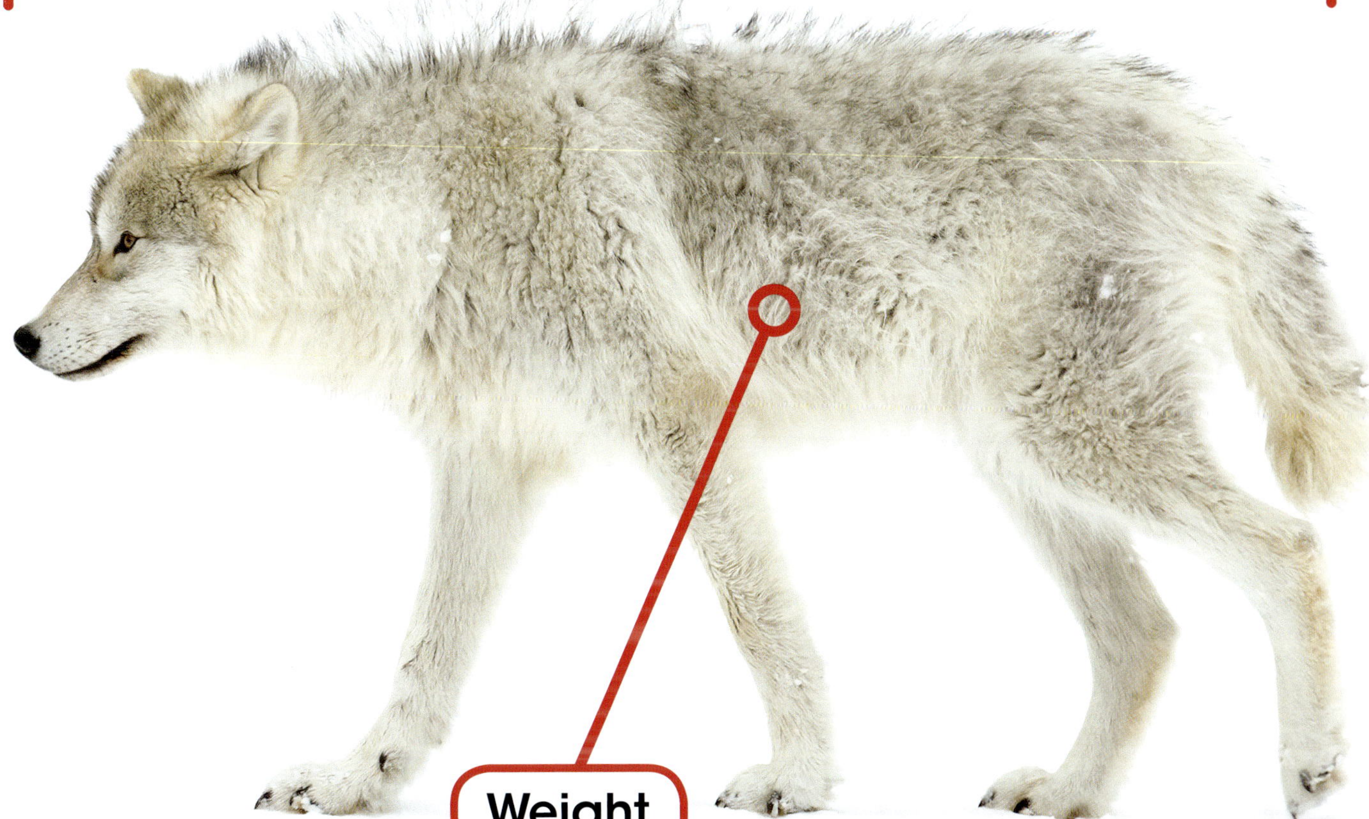

Weight

99 to 154.3 pounds (45 to 70 kg)

Pack Size

Two to twenty wolves, but seven or eight is most common

Territory

About 1,000 square miles (2,590 sq km) or more

GLOSSARY

adaptation: A change in a living thing that helps it live better in its habitat.

female: A girl.

male: A boy.

muzzle: The nose and mouth of an animal.

predator: An animal that hunts other animals for food.

prey: An animal hunted by another animal.

tundra: Cold northern lands that lack forests and have permanently frozen soil below the surface.

FOR MORE INFORMATION

WEBSITES

Arctic Wolf
animalia.bio/arctic-wolf
Color photographs highlight this website full of information on the Arctic wolf and its life.

White Wolf of the North
wwf.panda.org/our_work/wildlife/profiles/mammals/arcticwolf/
Learn more about helping Arctic wolves on this WWF website.

BOOKS

Brandenberg, Jim. *Face to Face with Wolves*. London, UK: Collins, 2017.

Flaherty, Williams. *Arctic Wolf*. Iqaluit, Nunavut; Toronto, Ontario: Inhabit Media, 2018.

Publisher's note to parents and teachers: Our editors have reviewed the websites listed here to make sure they're suitable for students. However, websites may change frequently. Please note that students should always be supervised when they access the internet.

INDEX